D0296158

OCEAN LIFE

OCEAN LIFE

Steve Parker

·PARRAGON·

Acknowledgements

Bruce Coleman Ltd/Quentin Bennett page 33; /**Mark N. Boulton** page 40; /**Fred Bruemmer** page 19; /**Jane Burton** pages 9, 30, 45, 66; /**Neville Coleman** page 67; /**Jules Cowan** page 6; /**Gerald Cubbitt** page 18; /**Adrian Davies** page 11; /**Francisco Erize** page 62; /**Jeff Foott** page 73; /**Francisco Futil** page 27; /**Michael Glover** pages 47, 60; /**Charles & Sandra Hood** page 43; /**Johnny Johnson** page 65; /**Luiz Claudio Marigo** page 69; /**Paul Meitz** page 14; /**M. Timothy O'Keefe** page 37; /**Dr Eckart Pott** pages 15, 25; /**Allan Power** pages 35, 48; /**Hans Reinhard** pages 13, 59; /**Carl Roessler** pages 49, 61; /**Dr Frieder Sauer** page 23; /**Nancy Sefton** pages 42, 44, 55; /**Kim Taylor** pages 26, 77; /**Bill Wood** pages 46, 51 left; /**WWF/Hal Whitehead** page 76; /**Christian Zuber** page 17.

The Image Bank/Derek Berwin page 58; /**Ocean Images** page 56.

Planet Earth Pictures/Richard Chesher page 51 right; /**Mary Clay** page 20; /**Mark Conlin** page 39 right; /**Peter David** pages 72, 75; /**Georgette Douvma** pages 28, 39 left; /**Jeannie Mackinnon** page 38; /**Peter Scoones** front cover, page 78; /**James D. Watt** page 52; /**Norbert Wu** pages 36, 70, 74.

First published in Great Britain in 1994 by
Parragon Book Service Ltd
Units 13-17, Avonbridge Trading Estate
Atlantic Road, Avonmouth
Bristol BS11 9QD

Publishing Manager: Sally Harper
Editor: Anne Crane
Design: Robert Mathias/Helen Mathias

ISBN 1 85813 851 5

Printed in Italy

Contents

CHAPTER ONE

At the Edge of the Sea

The seashore is one of the most varied, fast-changing and demanding of all habitats and is dominated by the tidal cycle. A few creatures, such as the fish and the more mobile crabs and prawns, move up and down with the water's edge, as the tide rises and falls, but many, such as sea snails and worms, stay in the same small area. Some, like mussels, barnacles and anemones, are permanently fixed in place, so they must be able to survive both in salt water when the tide is high, and in air when the tide recedes.

Under the water, oxygen can be obtained through gills, floating food particles are always available, and it rarely gets too hot or too cold, but salt levels are high, which has the effect of sucking vital water from an animal's body tissues.

Out of the water, oxygen is only available from the air and so lung-type body parts must take over from gills. Food needs to be hunted. The sun evaporates pure water from pools, so their salt concentration rises, then rain replaces salt water with fresh. This has the opposite effect and sucks out vital body salts and minerals.

The physical environment is just as changeable. Waves may pound the shore and batter everything in their path, churning the sand, hurling rocks, and uprooting plants and animals alike. Sea spray dries to leave thick encrustations of dehydrating salt. Animal's temperatures rise in the heat of the sun, then plummet as cold winds dry out their body tissues.

Despite all these difficulties, life thrives

FACING PAGE: *Around the shoreline, we can glimpse a small part of the immense and complex web of ocean life.*

in this narrow habitat and a huge variety of animals thrives on two main food sources. One is the bounty of the sea, for each incoming tide is cloudy with particles of nutritious food, brought in by the currents. Many animals strain these particles directly from the water, or suck them up from the sand and mud. The second source of food is the plants growing on the shore itself.

Time and tide

The Moon's gravity pulls the ocean's waters towards it, in a bulge that travels around the world as Moon and Earth spin in orbit. The result is the tides that creep in and out of the coastline approximately twice every day (in fact, high tides are some 12 hours 25 minutes apart). Every two weeks there are spring tides, when the Sun moves in line with the Moon. They drag the water even farther up the shore at high tide, and even farther down at low tide. These extreme high and low waters are interspersed with neap tides, when the difference between high and low tide is at its minimum. Superimposed on the tidal cycle are the effects of winds, storms, large waves and water currents.

Tides occur all around the world, but they are more pronounced in some places than others. In enclosed seas like the Mediterranean the difference between high and low tide is perhaps less than 1 m (39 in). In more exposed places, such as the west coast of Britain, the tidal range is much greater – 10 m (33 ft) or more in the Bristol Channel.

Tides wash up and down coasts of many different kinds. In polar regions, the coast may be the sheer ice cliff of a glacier, or the shallow slope of an ice shelf, but more familiar seasides are the sloping rocky and sandy shores, and sheer cliffs. Shingle shores have much larger particles, and mud flats far smaller ones. Each of these coastal environments supports its own community of plants and animals.

Zones on the rocky shore

Each animal and plant on the seashore has its favourite position, where it can survive best. Some prefer the high tide mark, where they rarely get wet, others are adapted to the almost continual submersion of the low water mark, while the

RIGHT: *Mid-shore boulders overgrown with filamentous algae and bladder wrack.*

more adaptable live in between.

At the lowest tide mark, the sublittoral fringe, animals and plants only poke above the surface as the tide turns. Huge seaweeds, such as oarweed and furbelows, form great forests, their root-like holdfasts grip the rocks and their leaf-like fronds wash to and fro in the waves, providing shelter for creatures such as sea urchins and blue-rayed limpets.

Sea squirts, such as *Aplidium* and *Pyura,* are strange bag-like creatures that live on some lower rocky shores. There are more than 2,000 species around the world. They usually exist in colonies, filtering food particles from the water. When the tide has left them uncovered, and when they are trodden on, they may squirt a jet of water into the air, hence their common name.

Between the tides

Animals and plants of the intertidal or eulittoral zone must be able to survive as well in the water as out of it. Seaweeds such as bladderwrack and knotted wrack partly dry out at low tide, but their tough, rubbery tissues suffer no ill effects. Bifurcaria weed only grows in rock pools, to avoid dehydration. Periwinkles *Littorina* are small sea snails which live between these sheltering seaweeds. When the tide is in, they wander over the rocks and scrape the thin algae from the surface with their file-like tongues. When the tide goes out, they hide inside their shells in rock crevices, to escape both their enemies and dehydration.

The shipworm *Teredo* is not a worm, but a worm-shaped mollusc with two small, clam-like 'shells' at one end. It uses these to rasp and rub with a rocking see-saw motion, in order to bore into driftwood – or the wood of ships or pier piles. The creature makes itself a long protective tunnel, from which it extends its two tube-like siphons to filter sea water for food.

Like the shipworm, the mussel *Mytilus* is a bivalve mollusc, that is, it has a shell in two parts or valves. It also survives by filtering food from the sea water. Mussels cling to the rocks in dense clusters on the lower shore, anchored by tough threads called byssus.

An ecosystem in miniature

As the tide goes out, some intertidal ani-

mals take refuge where water remains among the rock outcrops. This is the world of the rock pool, an ecosystem in miniature, with its own plants, predators and prey. Every tide changes the water, brings more food, and takes away waste.

In the rock pool, seaweeds such as the tough bladderwrack and delicate sea lettuce provide grazing for limpets. These are molluscs, relatives of the snails, with a single shell shaped like a flattened cone. They grip the surface with the powerful muscles of their broad, flat, fleshy 'foot', and not even the biggest wave can dislodge them. As the tide comes in, the limpet wanders slowly over the rock, grazing the thin covering of algae with its radula tongue. It only ventures around a small area, and usually returns to the same spot when the tide goes out. At its base, the limpet sucks itself onto the rock, fitting exactly into a little pit that its shell edge has worn away, so that it can cling even more tightly.

Yet another rocky-shore bivalve mollusc

LEFT: *This rock pool contains both brown and green seaweeds.*

is the piddock, which can actually bore a burrow right into solid rock, for ultimate protection. It feeds like the shipworm, extending two tube-shaped siphons into the water to suck in nutrient particles. The piddock can enlarge the inner part of its burrow as it grows, but this means it can never escape from its rocky hole, which is both home and tomb.

Rock-pool fish, such as shannies, gobies and blennies, are mostly small, alert and tough-skinned; they rely on their mottled brown colouring to hide among the weeds and pebbles at the bottom of the pool. Their eyes are almost on the top of their heads, to watch for predators such as gulls approaching from above. Gobies have their pelvic fins formed into a sucker, so they can fasten themselves to the rock surface in heavy seas.

The acorn barnacle is one of the most curious shore creatures. It is not a mollusc but a crustacean, a cousin of shrimps, crabs and lobsters. After a short period as a tiny drifting larva, it chooses a suitable rock surface and cements itself securely, by the antennae on its head. It then grows protective plates that resemble a miniature volcano, with a hole in the top. The hole can be sealed by a hard shelly 'door' when the tide is out. As the sea returns, the barnacle opens its door and sticks out feathery structures that are actually its legs. These beat through the water, gathering food particles. Hence the naturalists' description of the barnacle as an animal that 'glues its head to the rock and kicks food into its mouth'.

Ragworms hide under stones, waiting for other animals to wander past; up to 1 m (39 in) long, they are related to the familiar earthworms – but ragworms are fierce hunters with sharp teeth.

The dog whelk, another sea snail, is also a hunter. It has an extendible proboscis with a hard tip, which it uses to drill through the shells of mussels, barnacles, clams and limpets; it then sucks the victim's body out through the hole. The larger common whelk feeds mainly on carrion.

Yet another unlikely hunter is the beadlet sea anemone. This catches small fish, shrimps and other creatures with its stinging tentacles, and pulls them down to the stomach inside its 'stalk'. The sea hare, a molluscan relative of the snail, feeds on a diet of anemone tentacles.

Each habitat has its detritivores, feeding on leftovers and carrion and other debris. Shore crabs and hermit crabs clean up after the carnage, tearing up dead bodies into tiny pieces. Even some of the shells are re-used. When a growing hermit crab needs a new home, it selects a larger shell, usually from a whelk, to protect its soft rear body.

The high life

Above the mark of the highest spring tide is the splash zone, or littoral fringe, where only the occasional salty drenching affects life. Lichens, which are half-plant, half-fungus combinations, slowly encrust the rocks with patches of colour. Channelled wrack is a seaweed that lives along the lower edge of the littoral fringe; it is regularly dried to dark crackles by the wind, then drenched and rejuvenated by the biggest waves. Even some barnacles exist here. When the highest waves reach them, they wave their feathery legs in the spray, to grasp tiny food particles.

High above the rocky shore, cliffs pro-

RIGHT: *Marine iguanas sun themselves on rocky ledges by the sea.*

vide a home for birds that live off the sea. Gannets jostle for ledges on isolated stacks and rocky outcrops, where their eggs and babies are safe from land-based predators. Guillemots lay their eggs on crowded ledges alongside precarious sprigs of yellow sea cabbage. Puffins make burrows in the earth of the slopes, among pink clumps of sea campion, and some penguins breed on the bare ice of floating bergs and floes. All these birds feed on fish, squid and other sea creatures.

Sandy shores – buried life

Sandy beaches provide a less stable home than a rocky shore. There is no solid base, to which seaweeds and encrusting animals can cling. Each tide lifts and swirls the surface layers of sand and food particles and only animals that burrow beneath the surface can make homes here. Most of them live near the low-tide mark, where the sand rarely dries out.

Cockles, tellins, clams and razorshells are among the bivalve molluscs that burrow in sand and mud. They pull themselves under the sand with a muscular foot, and extend two long siphons up to the surface. Water is sucked into one siphon, the oxygen extracted and food filtered from it, body wastes are added, and the water is then squirted from the other siphon. The long shell valves of the razorshell resemble an old-fashioned cut-throat razor, hence its name. This streamlined bivalve is an extremely fast burrower, by pushing its foot down through the sand, widening it, pulling the shell down behind, then repeating the process, it can dig into the sand faster than you could.

Many kinds of worms live buried in the sand. The lugworm lives in a U-shaped burrow. Its squiggly wormcast of eaten sand is on the surface at one end, and a small conical depression marks the other, where it swallows more sand. On some shores, forests of small, stiff, sandy tubes stick up above the surface at the low-tide mark. These are made by peacock worms, as extensions to their burrows below the surface. Peacock worms filter food from the water with their branching, delicate 'peacock's tail' of tentacles, which protrude from the turret. At the slightest hint of danger, the worm whisks its tentacles deep into its protective tube.

Sand eels are fish, but not true eels; they are more closely related to herring.

ABOVE: *The common starfish hunts bivalve molluscs and other prey on sandy and rocky shores.*

FACING PAGE: *A large group of guillemots lay their eggs on crowded ledges.*

They swim in shiny shoals just above the sandy bottom, feeding on any smaller creatures they can find. Their pencil shape helps them to burrow into the sand, to hide from their many predators. Also hidden in the sand is the weever fish, with its poisonous back spines ready to sting any predatory flatfish that comes snuffling past.

The common starfish is an unlikely carnivore. It hunts bivalve molluscs and other prey on sandy and rocky shores. The starfish straddles its victim and relentlessly pulls apart its shell with its arms, which grip with hundreds of tiny sucker-tipped tube feet. The starfish then turns its stomach inside out through its mouth, pokes it into the shell of the prey, and digests and absorbs the flesh within.

Starfish belong to the large animal group called echinoderms, or 'spiny-skins'. Echinoderms live in the sea and have a circular body plan, rather than the usual two-sided, mirror-image design. Sea urchins are also echinoderms, and some types, including the sand and heart urchins, live buried in sandy shores.

All these animals hiding in the sand provide a rich food source for birds, who are able to probe with their long bills. Wading birds gather at low water to feed, but they are not all hunting for the same food. Dunlins and plovers have short beaks for delving just below the surface, mainly for tellins and shore crabs; oyster-catchers have medium-length bills and feed on cockles as well as oysters, while curlews have the longest bills and can reach down to lugworms.

High and dry

The high-water mark on a sandy beach is identified by a strand line – objects washed in by the tide and stranded high and dry. Beachcombers who push aside the plastic rubbish and dried seaweed seething with tiny sandhoppers, will find all manner of things. There are beautifully sculptured pieces of driftwood holed by the dreaded shipworm, shells of bivalve molluscs neatly drilled by the dog whelk, the ball-shaped cases of sea urchins; assorted fish skulls and bones, the single flattened oval 'bone' of the cuttlefish, clusters of whelk eggs like small knobbly sponges, and 'mermaid's purses', actually the empty egg cases of dogfish and rays.

Above the high-tide mark, the wind

RIGHT: *Ghost crabs take their name from their camouflage colours and nightly excursions.*

may blow sand from the beach to form sand dunes. Dune life is hard for the sand is dry, abrasive and shifting, and contains few nutrients. Plants such as sea holly and sea bindweed have tough leathery leaves, deep roots and long creeping stems, to try and stabilize their own patch. Marram grass, the most successful colonizer of dunes, has a vast root system that spreads wide and deep, its stems growing up through the shifting sand to form new clumps. When the weather is dry, it rolls up its tough leaves so they do not dry out. These resilient plants provide homes for tiny spiders and beetles, and flies and moths. Banded snails can survive by sealing themselves inside their shells during dry times.

No seashore habitat would be complete without its crabs, and the dunes are no

LEFT: *A leatherback turtle comes on to Rantau Abang Beach in the Malay Peninsula to lay its eggs.*

exception. The incredibly hardy shore crabs may venture here.

The problems on shingle beaches are even greater. The pebbles and boulders are slowly being ground down into sand, by the pounding waves, but as the shingle shifts with each wave, no animals or seaweeds can live here. Some hardy plants, such as sea blite and sea kale, can survive living above the high-water mark. The ringed plover even chooses the above-tide stones as a safe site to lay its eggs, each looking exactly like a pebble!

LEFT: *Male horseshoe crabs crowd atop an egg-laying female.*

Visitors to the beach

Beaches around the world provide temporary homes for many animals. Female sea turtles, such as the leatherback, haul themselves up their traditional breeding beaches, to dig holes for their eggs. After their labours, they return to the sea, leaving their babies to hatch weeks later, and scurry down to the waves on their own.

Female king crabs (horseshoe crabs) also come ashore to bury their eggs in a hole. But each drags a male with her, to fertilize the eggs. These strange armoured 'living fossils' are not true crabs but are more closely related to arachnids such as spiders and scorpions. These crabs usually stay partly buried on the sea bed, following the lifestyle of a carnivore-scavenger.

Some beautiful tropical beaches are made not of sand, but of millions of tiny fragments of shells. Many plants found on tropical islands, such as the coconut palm, shed their seeds onto the beach. These are caught by the tide and drift across the ocean, floating in their fibrous husks. They are finally washed up on far-away beaches, where if conditions are right, they germinate into more palm trees fringing the beach.

Chapter Two

Estuaries and Mangrove Swamps

FACING PAGE: *The aerial roots on this mangrove tree can be clearly seen.*

Where a river flows into the sea, fresh water gradually becomes salty. The bigger the river, the longer this takes. The fresh water pouring from the mouth of the Amazon – which represents one-fifth of all the river water on Earth – dilutes the salty water of the Atlantic for more than 150 km (93 miles) out to sea. The river water may become cooler, too, as it leaves the relative shallows of its land-edged channel, and, as the river water slows down, the smallest particles floating in it begin to sink to the bottom.

These are the peculiar conditions of the river mouth or estuary, a curious habitat which is half-fresh and half-salty, dominated by tides and mud. At high tide, the salty water pushes back the river and floods onto the surroundings flats of silt, mud and sand. Salt marshes with their fleshy-leaved plants, such as glasswort, sea aster and sea purslane, are found almost at the land's edge, and the salty water trickles among them, up the myriad tiny creeks and inlets. Sea lavender and pink thrift add their colourful flowers in season. Then the tide drops, and the river takes over, as fresh water washes the lower-lying areas while the higher land dries in the sun.

The type of estuary or coastal wetland depends on several factors, such as the shape of the land, the size of the river, and the local tides and sea currents. Some rivers pour off steep land straight into the sea, almost without an estuary at all. In other cases, the river meanders slowly for kilometres among flats and marshes, protected from the force of the waves by sandbars or shingle spits, and the estuarine habitat covers hundreds of square kilometres.

Few, but many

Estuaries tend to be home to relatively few species of animals, because of the fluctuating conditions and the general absence of plants. Only a few strongly-rooted, strap-leaved plants, such as cord grass and eel grass, can get a hold in the shifting mud, so estuarine animals tend to be small, and to hide in the mud and under stones while the tide is out. However, the species which are found in estuaries, are often present in incredible numbers. There may be over 20,000 tiny hydrobia spireshells, a small sea snail, in an area of just 1 sq m (11 sq feet).

Some of the animals filter edible floating bits and pieces swept in on the tide, or swept down by the river, while others feed on the mud itself. It looks unappetizing, yet it is extremely rich in organic matter. Worms, molluscs and crustaceans live close together in this nutrient-rich, but often oxygen-poor, environment.

The sand gapers or soft-shelled clams are large bivalve molluscs with fleshy siphons 30 cm (11.8 in) or longer. They suck water down one tube in the siphon, extract the oxygen and edible fragments, and pump it back up the other tube.

Another large bivalve is the peppery furrowshell with a pale shell as big as your palm. It burrows 20 cm (7.9 in) down, yet still its two long siphon tubes can reach up from the burrow and probe around the muddy surface like elephant trunks, picking up bits of food.

Worms and more worms

The red-backed ragworm is adapted to withstand occasional immersion in almost fresh water – something few true sea-dwellers can tolerate. It has a snout-like proboscis armed with powerful toothed 'jaws' for seizing smaller creatures, and it burrows into the soft mud to hide when needs be.

The cirratulus worm is a plump, pale, fleshy creature which lives in mud rich in rotting algae. It has two sets of tentacles. One set is designed to absorb as much oxygen from the water as possible, since oxygen is in short supply due to the rotting and decomposition. The second set

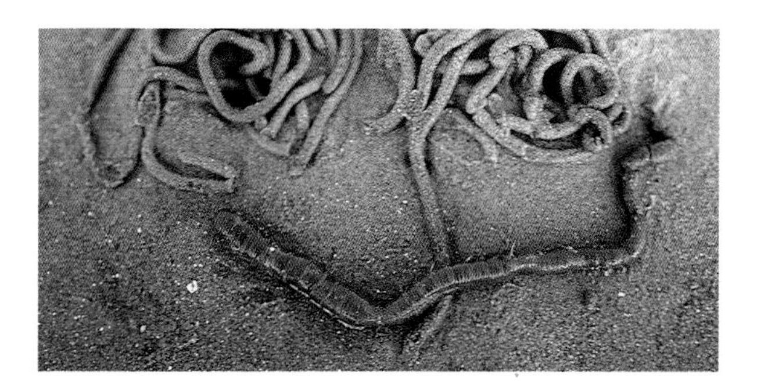

RIGHT: *The lugworm leaves wormcasts of eaten sand at one end of its burrow.*

traps particles of food and conveys them to the worm's mouth.

Of the many other worms in the estuary mud, perhaps the most extraordinary is the bootlace worm. It is brownish-black, thin and slimy – and as its name implies, very long. In fact, some specimens grow to over 30 m (98 ft) long, although 4 m (13 ft) is the average in most areas. This worm shelters under a stone in a tangled mass, preying on smaller animals or eating rotting debris. It watches with the numerous tiny, simple eyes along each side of its head for dark shapes that could mean danger.

Rich pickings

The small crustacean called corophium looks like a miniature hump-backed shrimp. It lives in a burrow, and builds a raised wall of mud around the entrance. At high tide it emerges and rakes the muddy surface with its many legs, to gather food.

A much larger crustacean is the blue swimmer crab; as crabs go, it is an excellent swimmer, using its rear legs which are flattened into paddle shapes.

Many of these mud-dwellers are active only at high tide. When the sea water covers the estuary mud and silt, their tentacles and siphons and legs and arms protrude and make the bottom look like a living deep-pile carpet. You can get some idea of their numbers by watching the wading birds such as curlews, knots and godwits, picking their way over the mud flats, jabbing with their beaks for creatures beneath. Plovers, lapwings and dunlins also probe in mud. Indeed, many estuaries are internationally important wildlife areas because they provide food for migrating birds who are passing through, or are wintering places for other bird migrants.

The estuarine catfish inhabits muddy estuaries and also sandy and rocky coasts; about 1 m (39 in) long, it has the typical 'cat's whiskers' barbels on its head, but a long, eel-like tail. By day it hides under a rock or stone, emerging at night to search for worms, shellfish and other prey.

Changeable chemistry and colour

Estuary residents may be limited in variety, but estuary visitors are certainly not. As the tide covers the flats, fish and crustaceans cruise just above the bottom, trying to detect life hidden just below. Many kinds of flatfish wander towards the river, but the estuarine specialist in this group is the flounder. It is at home in either fully salty sea water or fully fresh river water. It manages this feat by adjusting its body chemistry, and changing the amount of water it drinks (yes, fish do drink) and the volume of urine made by its kidneys. Flounders can travel kilometres up rivers, although they always return to the sea to spawn.

The flounder lies on its pale left side, and the upper or right side is olive-grey. Like most flatfish, it can vary the intensity of its coloration to blend in more accurately with the sea bed immediately around.

The grey mullets also have great salinity tolerance. They swim up river from estuaries, especially in summer. These mullets have a feeding method which is unusual among fish; they swallow the silt and mud, grind it to a smooth paste in the muscular stomach, and digest any nutrients, thus they fulfil the same role as earthworms.

Shocking jelly

Another flat fish (though not a flatfish) which may enter the estuary is the electric ray. Unlike many other rays, it is almost circular in outline when seen from above. Electric rays can stun victims or aggressors with powerful shocks from the jelly-like muscle 'batteries' along the sides of the body; a large ray can put out a shock of over 200 volts. These rays feed on molluscs, crustaceans and small fishes.

The estuary waters are a type of nursery for many fish. Some types of electric and other rays breed there. The voracious sea bass spawns in inshore waters, and the baby bass stay in the sheltered estuary for four or five years.

Other fish use the estuary as a crossroads, as they venture from fresh to salt water at a certain stage in their lives. After their 'childhood' in streams and fast rivers, salmon move out to sea for further growth, to attain maturity. A few years later they return to their birthplace, to spawn – and die. Eels do the opposite. They develop through the egg and larval stages at sea, head through the estuaries into fresh water to attain adulthood, them migrate back to the sea to breed (see page 63).

Among the mangroves

In some warmer coastal areas, the shape and nature of the shoreline, and the tides and currents, combine to produce the mangrove estuary or mangrove swamp. This is dominated by one or more species of mangroves, trees with stilt-like branching roots that hold the main trunk above the water surface. The mangroves slow the currents and encourage accumulation of mud and silt which harbours a rich selection of life. The mangrove winkle is a sea snail which slimes across the mud or climbs the mangrove roots. Some species of archer fish frequent mangrove lagoons, keeping an eye out for insects or other small animals just above the surface, on a leaf or twig or root. When the potential meal is spotted, the archer fish takes an extra-large mouthful of water and squirts it up and out into the air. The aim is to knock down the victim with this aquatic arrow, so the fish can snap it up.

ABOVE: *Commercial oyster beds in Brittany, France.*

One of the most entertaining of the mangrove swamp residents is the mudskipper. This smallish fish, on average not much bigger than your finger, has extra-large gill chambers which can hold a supply of water. The mudskipper absorbs the oxygen from this water, which it replenishes regularly by skittering to a small pool and taking a quick dip. It can

ABOVE: *A male fiddler crab signals with its brightly coloured claw to attract a mate.*

also absorb oxygen directly from the air through the blood-rich lining of its mouth and throat, so it is well fitted to stay out of water for considerable periods.

Some mudskippers feed by skimming up the tiny plants which live in the thin film of water on the surface. Others prey on drowning insects, small crabs and shrimps, worms and little fish. They set up territories at breeding time, each chasing away rivals from its own very desirable patch of mud.

Crabs with attitudes

Crabs with distinguishing features abound in estuary areas. The mangrove tree crab can climb trees, and often does, to escape water-bound predators or investigate possible food. Hordes of soldier crabs march across mud flats as though on army manoeuvres. Fiddler crabs signal to each other with overgrown pincers. The male has one very large claw, too big to be used for feeding, that he waves to and fro like a person playing the fiddle, in a complicated form of crab semaphore. These visual signals ward off other males from his territory, and also attract females.

Reptile and mammal

Larger animals are occasionally seen basking on the warm mud of tropical and subtropical estuaries. Undoubtedly the most fearsome is the estuarine or saltwater crocodile which, at a length of 7 m (23 ft) or more, it is the world's largest living reptile. Mainly nocturnal, it hunts turtles, birds, fish and mammals, and it can venture up rivers to pull victims from the bank, or swim far out to sea.

In great contrast to this sinister and ferocious creature are the docile, almost bumbling 'sea cows' or manatees. There are three species of these aquatic mammals, who frequent the coastal rivers, lagoons and estuaries of the tropical and subtropical Atlantic. They are the Florida and Antillean manatees in the west, and the West African manatee in the east. (A fourth species, the Amazon manatee, is strictly freshwater.) They are enormous beasts, 4 m (13 ft) long and 1,500 kg (3,307 lb) in weight. Strictly vegetarian, they swim lazily in the warm waters as they munch on submerged seagrasses and chomp on the plants lining estuaries, lagoons and riverbanks. Like any mammal, they breathe air and must sur-

LEFT: *The estuarine or saltwater crocodile has a good sense of smell out of the water, and its eyesight is excellent.*

face regularly. The manatees have whiskery, sagging snouts and large paddle-like forelimbs, but their hind limbs have disappeared, leaving a tail like a wide, flat, rounded flap.

Another sirenian (sea cow) species, the dugong, inhabits the warm coastal waters of the Indian and Pacific. It has a split tail resembling the flukes of a whale, and frequents estuaries but rarely ventures into fresh water.

The sirenians are slow and peaceful. This is not a recipe for success in the modern world. Despite being protected in many areas, they are hunted for food and for 'sport', and they are often injured by the propellers of powerboats or the keels of yachts.

CHAPTER THREE

Shallow Seas and Seabeds

The average depth of the sea, around our whole planet, is more than 3 km (1.9 miles), making most of the seas and seabeds extremely dark and deep. The shallow seas make up only a small fraction of the total area, mainly consisting of a narrow fringe around the edges of the world's continents and islands.

The sea's beds and bottoms form an area called the benthic habitat. In fact, there are several kinds of benthic subhabitats, such as sand, rock, mud, silt, pebble, and so on.

Despite its relatively small area, more animals and plants live in these shallow waters than in the deep sea. This is because, just as on land, and on the shore, and in the open ocean, life depends on sunlight.

Life-giving sunlight

On land, in fresh waters and in the seas, plants are at the base of the food chain. Their great service is to catch sunlight by special chemicals in their tissues, called photosynthetic pigments. These pigments turn the energy in sunlight into energy-rich nutrients, which the plants use to live and grow.

Animals eat the plants and take in the energy-rich substances to fuel their own life and growth processes. Other animals eat these animals, and the energy and nutrients are passed on. In this way the web of life builds up, and all of the energy used by living things can be traced back to original light energy that came from the Sun.

FACING PAGE: *Kelp forests provide homes for many other plants and a huge selection of animals.*

In the sea, the main plants are large seaweeds that live on the bottom, and microscopic plants that float in surface waters. All the animals in the sea, from crabs and worms to sharks and whales, depend in some way upon those plants.

Colourful plants

Light does not pass through water easily. As it gets deeper, it gets darker, so plant growth is limited to shallow water, where adequate light penetrates.

Seaweeds, or algae, come in different colours, depending on the light-absorbing properties of their photosynthetic pigments. There are three main groups: green, brown and red.

The different colours in white light are filtered out by sea water at different rates. This affects which seaweeds live where. In general, green seaweeds thrive where the light is brightest, at or near the surface. Brown seaweeds can live in deeper water. Red ones can grow even deeper, because

LEFT: *Flatfish such as plaice change the pattern and colour of their skin to match a different surface.*

their colour makes them more efficient at absorbing the small amounts of light which filter through the water.

From shore to shallow seabed

Some of the animals found in shallower water, perhaps 10–20 m (33–66 ft) deep, are similar to those living between the tides (see Chapter One). However, they are rarely exactly the same species; shore creatures are adapted to withstand exposure to the air, but not the continual immersion below the tide line.

Large brown seaweeds called kelps grow in underwater forests. Their long fronds wave in the currents and their large, knobbly holdfasts grip the rock. Like forests on land, the kelp forests provide homes for many other plants and a huge selection of animals. Around their bases grow other seaweeds, such as dulse. This red alga can cope with the shady conditions, in the same way that ferns grow under woodland trees.

Sea squirts live between the branched holdfasts, and green sea urchins wander between the stems. Sea mats coat the fronds higher up, like the pale, ghostly pile on an underwater carpet. These are colonies of tiny animals, protected in tiny cups or boxes which filter food from the water with their waving tentacles. They resemble miniature sea anemones or corals, but they are from a different animal group, the bryozoans.

Filtering for food

Where water currents are strong, they scour the seabed, removing the sand and mud, so the bottom is usually solid rock or pebbles. Less strong currents move the mud but leave the sand. In more settled waters, where the currents are gentle, the bottom is covered with mud or ooze which drifts down from above. Each of these seabed types provides different opportunities for animal life.

Filter-feeders strain nutritious particles such as plankton from the water. Many cannot move, so they usually live in places where the currents bring good supplies of such food. Fanworms, in their protective parchment-like tubes, extend their beautiful feathery tentacles into the current to do this. Brittlestars, which are slim-limbed, fast-moving relatives of starfish, hold their arms erect to do the same. They are mobile, and tend to collect in favourable currents. Barnacles are fixed to the rocks and trawl their feathery, filtering limbs through the water like their seashore cousins.

Sponges and sea squirts make their own currents, sucking in water and filtering it, before squirting it out. Sponges are the simplest of all multi-celled animals. They have no brain, nerves, heart or blood. When they die, their supporting skeletons of tiny silica shards, called spicules, remain and may find their way into the bathroom as a natural 'sponge'.

Deposits and detritus

Deposit-feeders live on what the sea deposits on its bottom – mainly sand, mud, silt and other ooze. It may not seem an appetizing menu, but the deposits contain tiny plants, bacteria and other microbes, and bits of detritus – dead animals and plants. Like the soil on land, it can be extremely nutritious and support a wide variety of life.

Crabs and lobsters are well-known detritus feeders. They nibble at any edible bits and pieces with their complex mouth parts. The great claws are used for tearing dead bodies apart (as well as for mating rituals). The lobster also has tiny pincers on its front feet, for picking up anything interesting and passing it to the mouth.

Under the sand

Few creatures live permanently on sandy seabeds. The surface is always shifting and there is nowhere to get a grip. But plenty of creatures walk or swim over it, looking for animals that live beneath it.

Garden eels are small, slim, colourful eels who populate certain patches of sand, forming colonies or 'gardens'. Each eel is less than 50 cm (20 in) long and lives tail-down in a corkscrew burrow. This is carefully constructed about 50 cm (20 in) away from the nearest neighbour, and so on. Hundreds of eels may stretch tens of metres across the sand. Just before daybreak, they pop their heads and necks out of their burrows, and spend the morning snapping up bits of food which drift past in the current. They look like the tall, brightly-coloured flowers of a cottage garden, waving in the breeze.

At noon the garden eels disappear as if by magic, down into their burrows for a rest. They reappear for the afternoon feeding session, then vanish as dusk falls. Their place is taken by fish such as rays, which swim just over the sandy surface, searching for the eels, shellfish and other creatures under the sand. A group of rays may arrive at dusk and 'mow the garden', feeding on any eels who do not pop back into their burrows quickly enough.

Side-faced flatfish

Flatfish such as flounders, sole, dab, plaice, turbot and halibut may look like rays. But they are very different kinds of fish.

Rays have skeletons not of bone but of rubbery cartilage. With sharks and skates, they form a large group called the cartilaginous fish, with some 710 species. A ray's body is flattened from top to bottom, so it lies on its underside on the sea bed.

Flatfish are ordinary fish, with the usual bony skeletons, like the other 20,000-odd bony fish species, but their bodies are flattened from side to side, so they lie on their sides on the seabed.

This change can be seen by following the flatfish's life cycle. Baby flatfish have an eye on each side of their head. As they

ABOVE: *The red mullet uses the sensitive 'whiskers' hanging from its lower lip to detect prey.*

grow, their bodies become thinner, and one eye moves around the head, next to the other. The mouth also turns sideways, apparently giving the fish a twisted smile. The underside is light, but the upper side is usually an intricate pattern of light and dark blotches, which blends exactly with the seabed.

Flatfish are the chameleons of the sea. They can change the pattern and colour of their skin when they move to a different surface, from tiny sand-coloured flecks to large dark- and light-brown patches. The change only takes a few minutes. The disguise is completed by a final flick of the fins to cover the edges of the body with sand.

Feeling for food

Flatfish hide on the seabed, to avoid being eaten by the skates and rays. Thornback rays glide just above the seabed hunting for buried creatures with their special electric sensory system. The ray generates an electric field around itself and can sense when the field is disturbed by another creature nearby, even if it is hiding out of sight.

Other fish feel for food hidden under the seabed in a more direct way. The red mullet has two barbels hanging from its lower lip which it trails along the surface of the sand to detect prey. The gurnard appears to walk over the surface with sensitive legs, but these are really the first few stiff, spiny fin rays of its pectoral fins. Both of these fish dig out and eat any small creatures that they find.

Buried danger

Fish, hunting over the sea-bed surface must beware, however, of the buried creature, that it will not be eaten, but do the eating. The wobbegong or carpet shark has excellent camouflage, with its mottled, knobbly body fringed with seaweedy lobes. It lies in wait for its food to swim close, and then opens its capacious mouth to gobble up its victim.

The shallow-water anglerfish adopts the same tactics but it also has a bait. This is a little 'worm' which is a fleshy lure that it waggles on a fin spine above its mouth. The lure is irresistible to hungry fish, and the anglerfish opens its mouth wide and sucks them in. Similar tactics are employed by its deep-sea relatives.

Another buried danger is the mantis

RIGHT: *With its mottled, knobbly body fringed with seaweedy lobes, the wobbegong shark has excellent camouflage.*

RIGHT: *A mantis shrimp poised ready to strike with its claws – one of the fastest movements in the animal kingdom.*

FACING PAGE: *This wrecked aircraft soon becomes a home for animals who like a firm foundation, such as corals, anemones and sponges.*

shrimp. It may not be very large, 10–20 cm (4–7.9 in) long, but it packs a big punch. This shrimp digs itself a burrow in the sand or mud, and hides in wait, then, when an unsuspecting crab or prawn wanders past, it darts out. Before the crab knows what has hit it, the shrimp thumps it with incredible speed and power using its spring-loaded claw. The crab's shell is smashed instantly, and the shrimp drags its dinner into its burrow.

The shipwreck scene

On a flat patch of sea bed wrecked ships provide an ideal home for animals who like a firm foundation. Very soon a wreck is draped with seaweeds and corals, anemones and sponges. The deadman's fingers and breadcrumb sponges waft water through tiny holes in the sides of their cylindrical bodies. Food particles are trapped by tiny hair-waving cells, and the

ABOVE: *The cuttlefish uses jet propulsion as an emergency mode of motion in addition to its undulating fins.*

used water exits through a larger, upper hole. Octopuses and conger eels take up residence in the crevices, hiding unseen until their prey wanders too close.

Giant groupers or sea bass are huge fish weighing 300 kg (661 lb) or more. They cruise through shallow waters past the wreck, picking off any smaller fish and nibbling at crabs. They have an appetite to match their weight, and some live for 75 years.

In a flap and a squirt

Many seabed animals walk, like crabs, or swim with their fins, like fish. However, some have more unorthodox ways of getting about. Scallops are bivalve molluscs related to clams and mussels which avoid their main predators, the starfish, by jet propulsion. The scallop senses the starfish with the simple eyes arranged like tiny jewels around the edges of its shell. These eyes cannot form a clear, detailed image, but they can sense patches of light and dark. As the starfish looms, the scallop suddenly snaps its shell halves together, expelling water so fast that it takes off. It then flaps its shells repeatedly to swim in a jerky, undulating fashion to a safer place.

Jet propulsion is also used by the cuttlefish, as a back-up mode of propulsion in addition to its undulating fins. These little animals, with their big eyes and long tentacles, are also molluscs, but related to squid and octopuses. They have a tube-

shaped siphon which they can use as a 'rocket thruster'. They point it the opposite way to the direction they want to go, squirt out water hard, and are propelled along.

Spiny lobsters are lobsters with spines instead of claws. They usually live hidden in rock crevices, coming out at night to forage alone for food. But when the winter approaches and the water gets cold, they move in great groups. Hundreds of thousands of spiny lobsters set off in convoys, spine-to-tail, across the sand to warmer waters.

ABOVE: *Unlike other types of lobster, spiny lobsters have spines in the place of claws.*

ABOVE LEFT: *An edible crab searches for food in a kelp forest.*

CHAPTER 4

Coral Reefs

Of all the wild places on Earth, only tropical rainforests surpass coral reefs in terms of the numbers and variety of animal life. Literally thousands of species of fish, crabs, anemones, worms, starfish, urchins and other marine creatures jostle for living space and food supply, in a riot of colour and activity. These animals show every kind of partnership and relationship in nature, from the parasites who live on and harm their hosts, to the situation called symbiosis where both members benefit. So the coral reef is a biologist's dream, as well as a paradise for snorkellers and scuba-divers; yet the whole reef is based on tiny, rubbery-bodied creatures which look rather like miniature sea anemones. They have short, cylindrical, stalk-like bodies and waving, stinging tentacles. These are the coral animals themselves, the coral polyps.

A single coral polyp is an unremarkable beast. The smallest are almost microscopic, while the biggest, such as the mushroom corals, are as large as your hand. However, most polyps range in size from that of a rice grain to that of a kidney bean. The secret of the reef lies in three factors which accompany these small, simple animals – skeletons, numbers and varieties.

Corals are classified in the cnidarian or coelenterate group, which contains simple animals such as sea anemones, jellyfish and sea fans. The coral polyp's body is soft and jelly-like, so it builds a limestone 'skeleton' around itself, usually in the shape of a cup or bowl.

FACING PAGE: *Beautiful but deadly, the lionfish has spines that contain a toxin that is fatal to humans.*

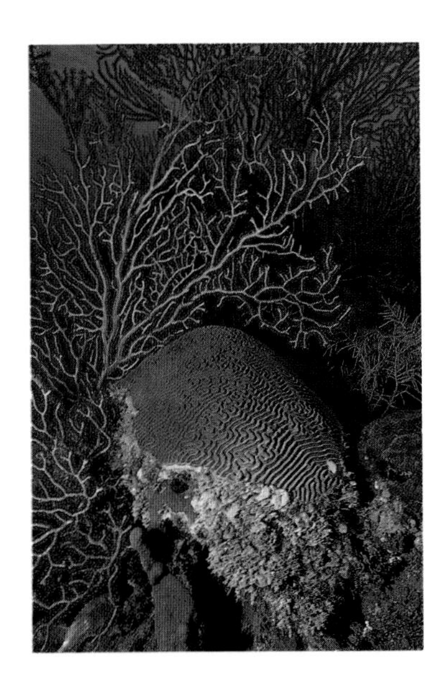

ABOVE: *Star and fan coral, growing together in the warm water of the Caribbean.*

Number and variety

One coral polyp in its stony cup-case may not look very impressive. But corals usually live in colonies, millions of them occupying millions of cups, which are joined together. When the polyps die, they leave their skeletons behind and new polyps settle on top and build their own rocky cups. Over hundreds and thousands of years, billions of skeletons accumulate, forming the rock of the reef.

There are hundreds of different species of coral and each produces a particular size and shape of stony skeleton, so its entire colony grows and develops in a certain way and pattern of branching. This is how the fantastic shapes build up, which give each type of coral its name, such as fan coral, brain coral, stag's-horn coral, mushroom coral, pillar coral, and many others.

Besides these hard corals, there are also hundreds of species of soft corals, which lack protective outer skeletons. Most of these look like small, slim sea anemones. They also live in colonies, carpeting the rocky surfaces of the reef. In fact, the polyps of a soft coral colony are often joined together at their lower ends, so that the polyp's stalk and tentacles project from a living blanket of communal jelly. This communal layer usually contains shared digestive cavities, tiny rods of mineral strengthening material called spicules, and bigger rods or tubes known as stolons.

How corals feed

Corals, like anemones, are flesh-eaters. They feed on tiny floating animals, called zooplankton, which they capture and sting with their tentacles. But corals have another food source as well, and this is linked to the reason why they grow only in certain seas.

The main coral reef areas are around the Caribbean, the Red Sea, parts of Africa, the Maldives and other islands in the Indian Ocean, South East Asia and north-east Australia, and Hawaii and certain other Pacific islands. The largest of all the reef complexes is Australia's Great Barrier Reef. It is some 2,300 km (1,429 miles) long and covers an area of 230,000 sq km (88,810 sq miles), almost three times the size of Scotland. It is by far the largest structure built by any life form (including ourselves) and can even be seen from the Moon.

The common factors all of these places share are warmth, clean and shallow waters and bright sunlight. Corals can grow in water at 16–35°C (61–95°F), but most reefs have formed in water with a much narrower range of 22–26°C (71.6–78.8°F). The local currents should bring enough floating food for the carnivorous polyps. The salinity, or saltiness, of the water should be about 35 parts per 1,000, or at least in the range 25–40 ppt. The water should be clear because silt or other fine sediments clog the polyps' tentacles and mouths, which is why few reefs are very near river outlets.

The water should also be clear because many coral polyps are like miniature greenhouses. Inside their body tissues are thousands of tiny single-celled plants, called zooxanthellae. These are algae, microscopic relatives of the larger seaweeds. They live naturally in the polyps. Like all plants, the zooxanthellae capture the energy in the bright sunlight by the process of photosynthesis, and use it to build up their own energy-rich nutrients and body tissues, from simple minerals that seep in through the polyp's body. During photosynthesis, they convert carbon dioxide and water into oxygen and energy-rich carbohydrates.

This animal-plant association is a fine example of symbiosis – a mutually helpful partnership between two very different species. The zooxanthellae inside the polyp have a safe place to live and a constant environment, rather than being exposed to the rigours of floating free in the ocean. The polyp gets a supply of oxygen for its own bodily process of respiration. It may also take up some of the nutrients that 'leak' out of the microscopic plants.

Helpful partners

The coral reef provides many other examples of symbiosis. Common clownfish, also called anemone fish, are gaily striped in orange and white. They rarely swim free, but live among the tentacles of large sea anemones (*Heteractis, Stoichactis* and *Dicosoma*). These tentacles would sting and paralyze other fish, but the clowns have a specialized coating of mucus over their bodies which resists the venom. Part of the mucus comes from the anemone itself, which uses the mucus to prevent its own tentacles considering

ABOVE: *The intricate patterns of brain coral, growing in the Red Sea.*

ABOVE: *Astonishing in its diversity of shapes and forms, coral grows only in warm, shallow waters.*

each other to be foreign objects and so stinging itself. The clownfish benefits from its safe haven, since few of its predators dare to encounter the tentacles. The anemone may eat bits of food dropped by its companions, as well as its normal diet of small fish, shrimps and tiny floating creatures. And the bright colours of the clownfishes may warn potential predators away from the anemone.

Cloak anemones are involved in a different animal–animal relationship. They take up residence on the shell of a hermit crab. The shell is not part of the hermit crab, since these crabs have soft, unprotected bodies but is usually the abandoned shell of a whelk or similar large mollusc, taken over by the hermit crab as its home. Hermit crabs scavenge for bits of food, and the anemone on its shell may benefit from catching the small animals that are stirred up, while the crab enjoys the protection of the anemone's stinging reputation. The red hairy hermit crab deliberately places a calliactus anemone onto its shell for this purpose. The anemone could move away on its stalk, if it wished, but does not. As the crab grows, it must search for a larger shell and transfer its soft body into the new home. Often, it pulls off and transfers the anemone as well.

Another highly evolved type of partnership involves at least fifty species of coral fish, mainly small wrasses and gobies. These are called cleaners. They feed by swimming or crawling over the bodies of larger animals, usually bigger fish such as wrasses and groupers, picking off and eating small parasites and pests, bits of fungus and other growths, fragments of dead skin and other debris. The client fish gets a free grooming and loses irritating parasites.

Some crustaceans are cleaners too, chiefly shrimps such as the red-backed shrimp of the Caribbean, and the banded coral shrimp found all over the world.

Cleaners have bright colours, such as the electric-blue body stripe of the cleaner wrasse, that act as an advertising sign, and each plies its trade at a particular site on the reef known as the cleaning station. Clients wait in an orderly fashion for the cleaner's attentions. The sight of a tiny cleaner wrasse or neon goby swimming around a huge grouper or tuskwrasse, even going into its mouth and gill chambers to pick off pests, is quite breathtaking.

RIGHT: *A clown fish darts between the tentacles of an anemone, immune from its stings.*

ABOVE: *Inside the mouth of a trout, this cleaner wrasse feeds off the bacteria clinging to its host.*

On the prowl

Not all coral reef relationships are so friendly. At the top of the food chains, and cruising close to the reef, are the predators. Among the sharks are the blacktip reef shark, the whitetip reef shark and the shortnose blacktip shark. These are relatively small shark species, seldom growing more than 1.5 m (5 ft) in length, and feed mainly on small crustaceans and fishes. Other sharks may come to the reef when young, or to breed, like the hammerhead.

The great barracuda grows up to 2 m (6.6 ft), and has a long mouth full of pointy teeth. It is a fast predator, alert and curious, and may approach close to human divers to investigate them. Barracudas have attacked people, but usually only when provoked or injured. As with sharks, the real risks are exaggerated.

Defence and disguise

Many fish rely on disguise to hide from both predators and prey. The stonefish lies still on the bottom, disguised as a weedy rock, waiting to gulp victims into its capacious mouth. The spines on its back can be erected to jab enemies with what is reputedly the mostly deadly venom of any fish.

Close behind in poison-power is the lionfish, also called the scorpionfish, with its long, colourful, lacy fins and fin spines. Some experts believe this spectacular decoration is to advertise the lionfish's deadly poison, which can be injected by a lash from one of the fin spines, others say the coloration is for camouflage. The fish hangs motionless among the seaweeds and corals, adjusting its coloration to merge into this wafting background. Small fish passing by do not see it until its too late. The lionfish can also spread its wing-like fins wide to herd prey into a tight corner, where it picks them off.

Boxfish have a physical defence rather than a chemical one. Their bodies are encased in bony plates of armour, making them very stiff. Since they are hardly the most lithe and flexible of swimmers, they have to waggle their fins and tails furiously to move at any speed.

The pufferfish has a surprise for predators who try to bite off more than they can chew. It can suck in water and inflate itself into a seawater balloon, far too big to swallow. The porcupinefish goes one better. It inflates its body, and as this happens, sharp prickles which normally lie flat on its skin become erect to upset the predator further.

Life and death on the reef

Moray eels of various colours and hues abound on coral reefs, hiding their leg-length bodies in holes and cracks, and waiting for their prey. The moray has a voracious appetite and will grab any fish, squid, cuttlefish or even skindiver that comes too near.

Coneshells are snail-like molluscs with the most intricate, beautifully coloured and patterned shells. They are also deadly, being among the most poisonous of all animals – especially the geographer's cone. The venom is employed to

ABOVE: *Cleverly disguised as a rock, the scorpionfish lurks, its spines ready to inject any unsuspecting passer-by.*

RIGHT: *Moray eels live in crevices in reefs and underwater cliffs, hiding unseen until lunch approaches.*

subdue prey, rather than repel enemies, although a coneshell will poison anything that disturbs it. The poison is in a small, hollow, dart-like spear than the coneshell shoots at its victim.

Fishes galore

Among the most colourful of the many dazzling fish are the many butterflyfish. Tall as they are long, their bodies are as thin as paper. The bright stripes and eyespots serve to break up the body outline and blend in with the coral backcloth, to fool predators into attacking the tail rather than the head, and also to advertise territorial ownership or a vacancy for a mate. The black and white angelfish, in particular, is hard to see among the shadows on the reef.

Surgeonfish have intensely bright colours, and also weapons in the shape of sharp spines on either side of their tails, which are shaped like the surgeon's scalpel. The blades fit into grooves until the fish is upset, and then they flick out as the fish swishes its tail. Triggerfish have a special hinged spine on the back, like the trigger of a gun, so they can wedge themselves into cracks when threatened.

Two other types of fish show features that are common to many reef-dwellers: sexual dimorphism, sex change and home ownership. Tiny jewelfish, like many other reef fishes, show sexual dimorphism. This means the males are different in size, shape or colour to the females. Male jewelfish are brightly coloured, with rounded fins and long dorsal fin spines, but the females are a more ordinary dull orange with angular fins.

As the jewelfish mature, they become females, then some change into males as they get older. This change of sex with age is a common occurrence among coral reef fish.

FACING PAGE: *The angel fish is one of the many colourful species that inhabit coral reefs.*

Damselfish are diligent home-owners. They mark out an area of the reef as their own, and defend it virtually to the death, driving off intruders with ferocious visual displays and loud clicking noises. Mated pairs clean their home thoroughly before they lay eggs.

Night on the reef

As dusk falls on the reef, the daytime creatures hide away and sleep. Clownfish nestle deep within their sea anemone tentacles, damselfish hide between the coral fronds and the Moorish idol's bright white and yellow pattern dulls to grey. Parrotfish find a crack and exude slime from their mucus glands over their bodies to make a jelly-like sleeping bag. The bag disguises the texture, shape and smell of the fish from night prowlers such as lionfish and bigeyes.

The rising plankton provides midnight snacks for the filter-feeders on the reef. The coral polyps, virtually inactive by day, come alive, protruding their tentacles from their stony boxes to grasp at the drifting particles. Brightly coloured sea-lilies climb up to wave their delicate arms in the rich soup. Colourful nocturnal sea slugs nose and browse among the corals, alongside the debris-picking shrimps. The sea slugs, or nudibranchs, are slug-shaped but they have elaborate fringes, frills and tassels, and come in the most vivid colours and patterns. The tassels on their backs are gills for absorbing oxygen from the water.

Sea slugs glide among the coral on their muscular feet, but most can swim quite well by undulating their frills. Some are no larger than a grain of sugar, while the Spanish dancer is more than 30 cm (11.8 in) long. All sea slugs are carnivores, rasping away with their file-like tongues at coral polyps, sea anemones, sponges and other nudibranchs. They are also efficient recyclers. Some house algal cells in their tissues, which they obtain from the coral polyps they have eaten, and which provide the slug with extra food from sunlight. Some re-use the stinging cells from the tentacles of the sea anemones that they eat, employing them for defence. Many sea slugs secrete poisonous slime as a deterrent, and some even produce sulphuric acid! Like most molluscs, each nudibranch possesses both male and female organs, and produces a large ribbon of eggs.

The flashlight fish has luminous patches under its eyes containing light-making bacteria. The fish can turn them on and off both to see in the dark and to lure prey fish near its mouth.

Invertebrate life

The coral reef is crammed with invertebrates, from the simplest sponges to the cleverest octopuses. These also have their day and night shifts. Sea anemones wave their tentacles. Giant clams filter the water and their simple 'eyes' around the frilly mantle edges watch for the dark shadow of a predator. Ragworms writhe among the rocks, seizing small animals in their tooth-studded probosces. Crabs, as ever, snip and pick among the debris for edible particles.

ABOVE: *The crown of thorns starfish is one of the major predators on coral reefs.*

ABOVE LEFT: *Unlike its terrestrial namesake, the carnivorous sea slug is flamboyant and beautiful.*

Chapter Five

The Open Ocean

The immense stretches of ocean may all look much the same to us, above the waves. But on and in the water, things are very different. There are distinct areas, each with its own water temperature, current pattern, nutrient levels and other features. The open ocean is a mosaic of these subtly differing habitats and the myriad animal species, from worms to whales, are distributed accordingly.

A stream of warm water

Sailors and fishermen have long understood the variety in sea conditions. One of the world's great ocean currents is the Gulf Stream, in the Atlantic Ocean. It is a body of warmer water that flows steadily from the Caribbean up the east coast of North America, and across to Europe. The Gulf Stream contains fewer nutrients than the surrounding water, and so supports less planktonic life, and therefore fewer larger animals in the oceanic food web – all the way up to the great whales.

Basis of the food webs

On land, we can see the trees, grasses and other plants on which all animal life depends, and which form the basis of the food webs. On the shore, we can point out the algae that fulfil the same ecological role. In the open ocean, the plants are still there, but they are much less obvious – they are microscopic phytoplankton.

The phytoplankton are tiny green cells, mostly diatoms and dinoflagellates. The

FACING PAGE: *One of the biggest fish of the open ocean, the blue marlin can grow up to 4 m (13 ft) in length.*

harvesters of these floating pastures are the miniature animals of the zooplankton. Many are the larvae of worms, starfish, sea anemones, crabs and fish. Should they survive, and the odds are millions to one against, they will continue to grow and develop into much larger adults. However, most of the organisms in the zooplankton are tiny but fully grown crustaceans. They are copepods, and they look like the little 'water fleas' of freshwater ponds, and they feed on the microphytoplankton. The gigantic swarms of billions of copepods, across the vast oceans, make them the most numerous animals on the planet.

Larger residents of the plankton are the finger-sized, shrimp-like animal called krill. They, too, live in huge quantities, and serve as food for seals, penguins and the biggest of animals, the rorquals or great whales. Krill feed on diatoms and other tiny components of the plant plankton. In a dense swarm there may be more than 50,000 krill in one cubic metre of sea water, and the swarm will stretch for many kilometres.

Carnivores in the plankton range from tiny single-celled animals such as foraminiferans and radiolarians, through to sea gooseberries, worms, arrow-worms, jellyfish and open-ocean molluscs such as squid.

Sieving with gills

Most links in the food chain follow the general path from smaller animals to larger ones. The phyto- and zooplankton are eaten by young fish and the larvae of animals such as crabs and starfish, and so on. At the summit of the food webs are spectacular predators, such as the swordfish, sailfish and marlin.

Certain fishes 'miss out' on the food chain links and consume the plankton directly, by filter-feeding. The herring eats not with its mouth, which has only small, feeble teeth, but with its gills, which are equipped with long comb-like sieves, called gill rakers. These collect tiny organisms from the water flowing past them, which the herring then swallows.

The gigantic shoals of herring, and schools of their close relatives the pilchards, menhaden and anchovies, are themselves preyed upon by larger fish, dolphins, seals and seabirds. They are very important links in the marine food chains.

RIGHT: *Small fish such as these fingerlings keep together for safety; any stragglers risk being picked off by predators.*

Confuse-a-predator

Planktonic organisms drift and are largely at the mercy of currents. They are mainly transparent to avoid their enemies. More active pelagic animals – open-ocean swimmers – have evolved different disguises. Mackerel live in large schools, feeding off plankton while they are young and larger animals as they get older. Like many fish of the surface waters, they use counter-shading as camouflage. The top of the body is rippled dark and light blue, the sides are silver-grey and the underside is almost white. The theory is that light always shines from above, illuminating the darker back and shading the paler belly, to make the two surfaces appear much the same. Together with their shiny reflective scales, this serves to break up

the outline of the fish, so a predator has less chance of making contact or getting a good grip.

Countershading serves equally well to camouflage or disguise predators as well as prey, so it is used by the large predatory fish as well. The tuna is countershaded dark blue above through to silver below, and has stripes running along its sides which add to the effect when it swims very fast.

This technique works well in open water. Where there are weeds to imitate, more elaborate disguises can be used. Great tangles of sargassum weed drift at the surface of the sea in the Sargasso Sea of the mid-Atlantic. These floating communities have their own unique animals, found nowhere else. The sargassum fish is related to the frogfish and both are masters of seabed disguise. The sargassum fish's body is splotched brown, yellow and white, and fringed with tassels and frills. In this way it hides from both its predators and prey.

LEFT: *A shark's acute sense of smell and keen hearing enable it to accurately pinpoint prey.*

Top fish

The supreme predators of the sea are superbly adapted for speed. Marlin grow to 4.5 m (14.8 ft) long. They use record-breaking swimming speeds to catch schooling fish, which they may swipe or stun or impale with their long, pointed noses, known as bills. A leaping blue marlin is one of the world's great wildlife sights, as the perfectly streamlined fish bursts from the surface and leaps to heights more than three times its body length.

Swordfish are also members of the bill-fish group. They grow even bigger, to 5 m (16.4 ft). Baby swordfish do not have a sword – their bill gets longer as its owner grows. No one is sure what the sword is used for, though some snapped-off swords have been found deeply embedded in the sides of wooden ships.

The large sharks, such as the great white, the mako and the hammerhead, which feed on large fishes, penguins, seals and dolphins are adapted in every detail as efficient killers. They have excellent senses of smell, touch, hearing and they can sense electrical impulses made by the active muscles of prey animals. They can also swim fast with the minimum of effort.

The largest predatory shark is the great white or 'man-eater'. There have been sensible sightings of great whites perhaps 9 m (29.5 ft) long, but the greatest measured and fully authenticated size is 6.4 m (21 ft) in length and just over 3 tonnes in weight. Next come the tiger shark, at 5–6 m (16.4–20 ft), and the Greenland or sleeper shark, which may rarely exceed 6 m (19.7 ft).

Playful killers

Dolphins are also one of the top predators. These fish-shaped mammals are small whales. They live in social groups, rolling and playing near the surface, 'spy hopping' by standing up straight half out of the water, and 'porpoising' by leaping out and then going under again at speed, especially in the bow waves of boats. Dolphins dive to catch fishes and squid, sometimes at depths exceeding 200 m (656 ft), then surface to breathe fresh air again. They find their prey in the gloom by a sound-radar system. The dolphin emits high-pitched clicks and squeaks, listens to the pattern and time delay of the

returning echoes, and analyzes the sizes and positions of objects nearby. The system is called echolocation, and is used in a parallel way by night-flying bats.

Seals are also mammals, and have thick layers of fat and fur to keep them warm. Few seals are found in the open ocean; most are coastal animals and haul out on land to rest. The harp seals probably spend most of their time in the sea, making very deep dives to feed on crustaceans. The pups are born on ice floes and are deserted by their mothers after less than a month's suckling, forcing the hungry youngsters to take to the water. The champion diver among seals is the Weddell seal, which has been tracked down to around 700 m (2,297 ft), but may be able to go even deeper.

RIGHT: *Antarctica cannot support animal life permanently, but in the spring, harp seals come ashore to breed.*

FACING PAGE: *Dolphins use sonar to track other dolphins or schools of fish, and to monitor their own location.*

Giants of the seas

True ocean giants can maintain their great bodies only with a reliable, constant food supply – the plankton. The ocean sunfish grows to 4 m (13 ft) long, from mouth to tail, and 6 m (19.7 ft) 'tall' from dorsal fin-tip to anal fin-tip, yet this massive fish has a small beaked mouth and feeds on tiny planktonic jellyfish.

One of the biggest fish in the sea is the basking shark, at 12 m (39.4 ft) or longer. Even larger is the greatest of all fish, the whale shark, which may exceed 16 m (52.5 ft). Both of these monsters feed by peacefully cruising through the plankton with their huge mouths open. They filter enormous volumes of water for small tiny planktonic organisms, which are trapped by bristly combs on the gills.

A close cousin of the sharks is the massive manta ray, or giant devilfish. It may be more than 5 m (16.25 ft) long and nearly 7 m (23 ft) wide across its wing-tips, and is a terrifying spectacle as it

LEFT: *This photograph of a basking shark's mouth shows the plankton on which the shark feeds.*

lazily flaps its wings (pectoral fins) and 'flies' lazily through the water, like a monstrous underwater bat. But again, it feeds only on the tiniest animals.

Even bigger are the largest marine mammals – great whales such as the humpback, right, blue, fin, sei and minke. They have huge mouths lined with strips of a horny substance, called baleen or whalebone, which hang down like great tall teeth. The baleen plates have frilly fringes that strain krill and other small animals.

The right whales have the longest baleen strips, which measure up to 4 m (13 ft). They got their name because they are slow, and easy to spot, and in the days of whaling they were considered the 'right' whales to catch. The blue whale is probably the largest animal that has ever lived on Earth. It reached 30 m (98 ft) in length and over 170 tonnes in weight, before commercial whaling decimated the species.

Sea otters are not truly oceanic, like the

RIGHT: *Manta ray swim just over the sea bed, searching for eels, shellfishes and other creatures under the sand.*

whales, for they stay close to shore, but even so, they spend most of their lives in the water and rarely come onto land. The sea otter keeps warm by a layer of air trapped in its thick fur; it has no layer of fatty blubber under its skin like most other sea mammals. It feeds by diving to pick up shellfish from the seabed and bringing them to the surface, together with a large pebble. The otter then lies on its back in the water, places the stone on its chest, and hammers the shellfish against it. At night the otter wraps itself in the fronds of seaweeds, so it will not drift while asleep.

Epic journeys

At different times of year, some parts of the oceans are more productive than others, so many animals migrate huge distances. Among the best known is the salmon. A typical salmon spends five or six years at sea, feeding on small crustaceans, while growing into a mature adult. Then it sets off back through the estuary and up the river, to the place where it was spawned up to ten years before. How salmon find their home rivers from a starting point thousands of kilometres out to sea, is still something of a puzzle. However it is known that this fish is extraordinarily sensitive to the 'smell' of the chemical profile contained in water from its home stream. It can also detect the Earth's natural magnetism, and the direction and strength of the ocean currents. The adult salmon change into their bright breeding colours as they struggle upstream, against waterfalls, dams and rapids, to reach their goal where they spawn and die.

For centuries, another great marine puzzle was the birth of the eel. No adult mature European eel, ready to spawn, had ever been found.

Now we know the answer. Eels live most of their lives in rivers, but on reaching adult size, they head for the open sea. They may even wriggle across land, if they have to. On arriving at the estuary, they swim 5,000 km (3,107 miles) across the Atlantic to the Sargasso Sea. When they arrive, they spawn in deep water, and die. The eggs develop into tiny,

LEFT: *Right whales have horny plates in their mouths instead of teeth, to strain minute krill from the water.*

transparent, leaf-shaped larvae which spend three years floating back to European waters in the surface currents. By the time they arrive, they are the mini-eels called elvers. These swim back up the rivers, and so the cycle is complete.

Around the world

The record-breaking migrator is the Arctic tern. It flies from its Arctic breeding grounds to the Antarctic, to avoid the long, dark northern winter. Then it flies back again to avoid the long, dark southern winter. These journeys can be 15,000 km (9,320 miles) – each way. While at sea, these graceful and delicate little birds feed on fish, squid and crustaceans, plunge-diving into the water to catch them.

The tern may travel the greatest distance but the most air miles may be notched up by the wandering albatross, which has the longest wings of any bird, over 3 m (9.8 ft) from tip to tip. Strong and narrow, these wings are ideal for gliding and soaring, obtaining lift from the sea breezes and rising warm air currents. These huge birds alight only at breeding time, mainly on isolated islands in the southern seas. Each female lays only one egg, which takes nearly three months to hatch, and which is then fed by its parents for nearly a year.

Terns and albatrosses are good at gliding, but the best stunt-flier is probably the gannet. It flaps slowly above the waves, looking for prey, and then dives like an arrow from 30 m (98 ft) or higher. The gannet has a specially strong skull to absorb the impact of the hitting the water.

Flying or swimming?

Some seabirds are better at swimming than flying. Cormorants feed on fish which they catch by diving. They propel themselves by paddling with their webbed feet, and steer with their tails. One species, the flightless cormorant from the Galapagos islands, has lost its power of flight altogether.

Penguins have also chosen to live beneath the waves rather than above them. They too can no longer fly, but they are excellent swimmers. Like albatrosses and terns, they leave the sea only to breed; the rest of their time is spent chasing fish and other animals under the water. Some penguins can dive for five

FACING PAGE: *The black-browed albatross is an inhabitant of the waters around the Antarctic.*

minutes and to depths of 200 m (656 ft), before they must surface for air.

Flying fish actually fly better than penguins. They have very long pectoral fins, which they extend after leaping from the water at speed, and can glide for well over 200 m (656 ft), at a height of 2–3 m (6.6–9.8 ft), if they catch the wind correctly. Flying fish take to the air to escape when they are being chased by speedy predators such as tuna.

Floaters

Some pelagic animals have specially designed floats to stop them sinking. The by-the-wind-sailor is a type of upside-down sea anemone, which trails stinging tentacles that hang down from an air-filled bag made of an almost plastic-like substance. The bag has a sail running diagonally across it, so the animal can be blown by the wind. The man 'o' war is not one jellyfish but a whole colony of anemone-like creatures, hanging upside

LEFT: *Like other penguins, the jackass penguin cannot fly but it is a strong swimmer, using its wings as flippers.*

and each doing a different job. Some form the gas sac, some trail stinging tentacles over 10 m (32.8 ft) long, while others have mouths to take in the food, or muscles to swim, or are responsible for reproduction. The gas in the sac can be let out, so the whole man 'o' war colony can sink in bad weather. Every so often, too, the sac flips over to keep itself moist.

The floating sea slug feeds mainly by nibbling the man 'o' war, being immune to the stinging tentacles. It has feathery 'wings' on the sides of its body to stop it sinking. Another mollusc, the violet sea snail, floats by making itself a frothy raft of bubbles.

Some pelagic animals use artificial buoyancy aids. Goose barnacles are stalked crustaceans, relatives of the rocky-shore barnacles (see page 12), which sweep food particles from the water with their feathery legs. Like their shore cousins, they need something to cling to, so early in life, when as tiny larval barnacles they prepare to settle and stick, they

RIGHT: *The slender suckerfish attaches itself to a shark and feeds on the bacteria on its skin without harming its host.*

may choose anything that floats, such as a buoy, cable, net or boat. Huge colonies of goose barnacles on the hull of a boat seriously reduce the speed of the vessel and increase the use of fuel.

Yet another way of travelling across the open ocean is to hitch a lift. Remora fish are also called sharksuckers, because their favourite mode of transportation is to stick to a large shark. The remora has a sucker pad on top of its head, made from modified fins. When attached, it may slow the shark slightly, but otherwise it does its host no harm. Indeed the remora may feed on the shark's parasites, occasionally making a brief excursion to catch larger prey or snatch some of the leftovers from the shark's meal.

Turning turtle

Surprisingly, the open ocean has its fair share of reptiles, too. Sea turtles swim with their long front flippers and steering with their short hind flippers. They find moving about on land very difficult, but the females haul themselves onto sandy beaches to dig holes and lay their eggs.

Green turtles feed on sea grass and seaweed which do not grow near their nesting sites in the mid-Atlantic. So each turtle migrates to the beach of its birth, covering many hundreds of miles, to lay its eggs. Leatherback turtles feed on the jellyfish that float with the plankton. Loggerhead turtles feed on crustaceans, molluscs and sponges which they chomp with their powerful beak-like jaws.

Sea snakes spend virtually their entire lives at sea, feeding on small fishes. The sea snake's body is flattened from side-to-side so it swims fairly well, but it must be sure that it can subdue its prey quickly, to avoid losing it, so when it bites, this snake injects one of the strongest poisons known in the animal kingdom: the victim dies almost instantly. Sea snakes breathe air, like their land relations, but some species can stay under the water for two hours. Their muscles are adapted solely for swimming, so that on the shore, they are almost helpless. They have gone one better than the sea turtles since they even breed at sea, producing fully-formed live offspring. Sailors have reported great 'rafts' many kilometres long of writhing sea snakes, which contain countless individuals and may be breeding groups – yet another fascinating surprise of the open ocean.

FACING PAGE: *After coming ashore to lay its eggs, the loggerhead turtle returns to the sea, leaving their babies to hatch weeks later.*

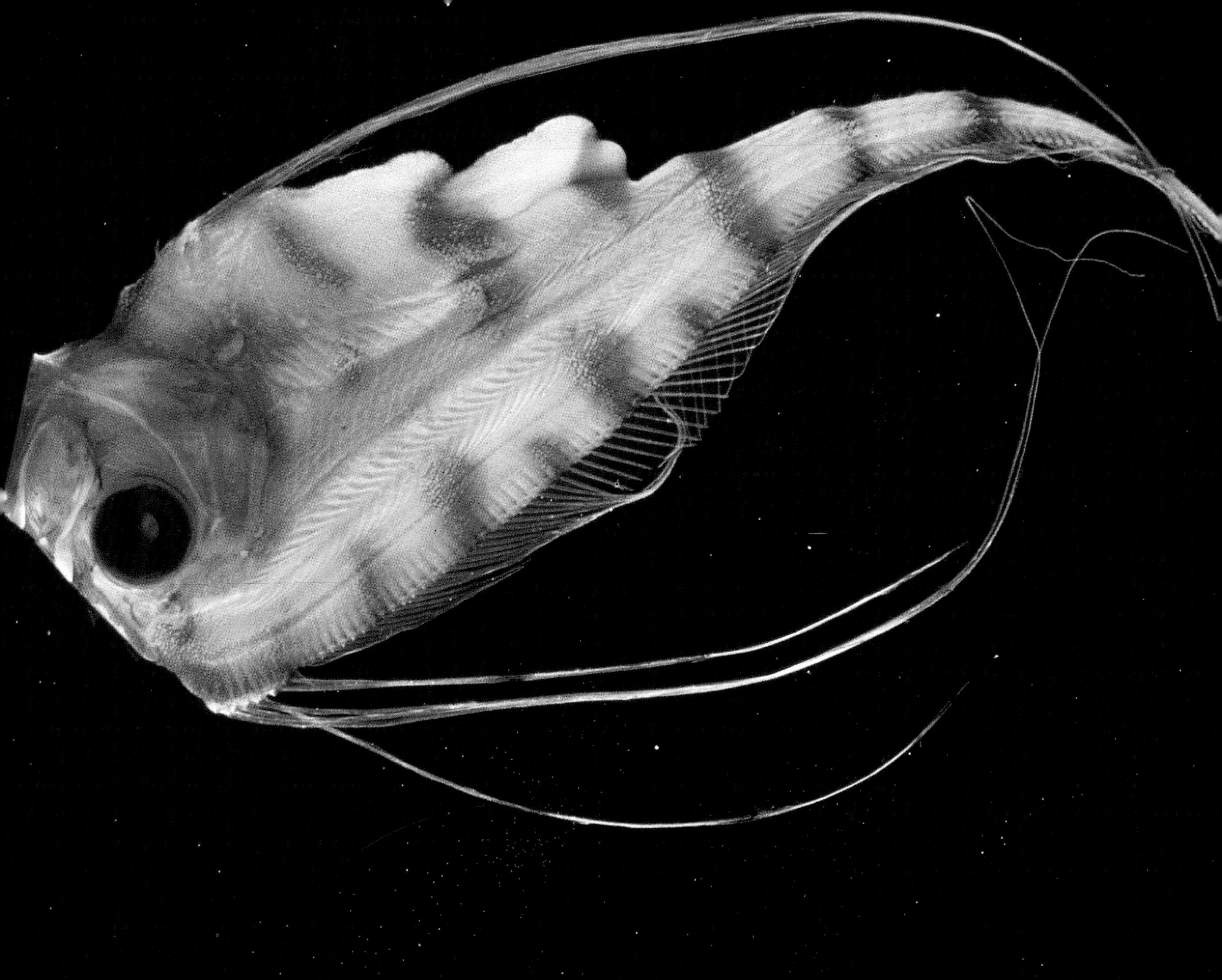

CHAPTER SIX

The Darkest Depths

FACING PAGE: *The subject of many sea-serpent myths, the crested oarfish grows up to 12.2 m (40 ft) in length.*

Living things become rarer as the water gets deeper, because life gets tougher. Some eighty-six per cent of the seas covering our planet are more than 1,000 m (3,280 ft) deep. The sun's light and warmth have never penetrated these depths, where the temperature is just a few degrees above freezing. Without sunlight, no plants can grow, so the food chains of the animals in this cold, silent, inky-black world depend on nutrients drifting down from the surface layers.

The deepest ocean is called the abyssal zone. Despite the darkness, the cold, the lack of food and the immense pressure of thousands of kilometres of water above, animals do exist. They are some of the strangest creatures on our planet, adapted to conditions we can hardly imagine.

The nature of the deep seabed

The ancient Greeks suspected that the seas were incredibly deep and populated by monsters, but they had no proof and it was not until this century, that both beliefs were shown to be true.

Sonar equipment uses sound beamed out from an underwater loudspeaker. The sound waves bounce off objects and return to an underwater microphone called a hydrophone. This detects the echoes, which are analyzed by computer to see how far the sounds have travelled. Sonar has revealed the contours of the

ABOVE: *A number of tiny, parasitic male anglerfish can be seen attached to this female anglerfish.*

deepest seabeds. But even today, relatively little is known about what is by far the world's largest habitat. Only the occasional bizarre catch by deep-sea trawler boats, and deep dives by specialized bathyspheres, remote-controlled submarines and other craft, give a glimpse into this mysterious world.

The deep seabed is made up of mountain chains, immense flat ooze-covered plateaux, and cliff-sided canyons or trenches. These features are all much bigger than the same ones on land. The deepest ocean valley, the Marianas Trench in the north-western Pacific Ocean, plunges over 11 km (7 miles) below the surface.

Always raining

The main source of food is a persistent light rain of particles from above. It includes dead planktonic organisms, scraps of meat from surface predators munching on prey, and droppings galore. These carpet the seabed, providing a wonderful feast for detritus feeders like sea cucumbers.

Sea cucumbers are echinoderms, related to starfish and sea urchins. They live on most sea bottoms, but they come into their own in deep waters where they sweep the muddy particles from the seabed with frilly tentacles arranged around the mouth. Some flattened sea cucumbers can swim, and cruise the slow deepsea currents, filtering out food particles.

Finding a mate

A significant problem for deep-sea animals is finding a mate. The chance of bumping into another animal of the same species but the other sex, in such vast blackness, is very slim. When a male deep-sea anglerfish finds a female, he bites her – and never lets go! His mouth grows onto her skin, his body merges into hers and he becomes part of the female. Almost all of his internal organs disappear, apart from the testes which are needed for breeding. The male spends the rest of his life as a shrunken parasite on his mate, useful only for providing the sperm which fertilize her eggs at breeding time.

Brittlestars and some sea cucumbers have solved the problem of finding a mate in another way. They have become hermaphrodite – both male and female at the same time. Whenever any two brittlestars meet, they can mate, since they are both guaranteed to be members of the opposite sex.

Brittlestars also give their offspring a better chance of survival, by keeping them inside their bodies until they have developed into tiny adults, so they can start sweeping the deep-sea floor as soon as they are born. These creatures are echinoderms; although they resemble starfish, they are more closely related to sea cucumbers. The deep-sea species live like their shallow-water cousins (see page 16). They are detritus feeders, picking up bits of food with rows of sticky tube feet along their arms. The tube feet pass the food particles to a channel in each arm, lined with tiny beating hairs. The hairs swish the food along the arms to the mouth in the middle of the brittlestar's body.

ABOVE: *Brittlestars are slim-limbed, fast-moving relatives of starfishes.*

Light in the darkness

Detritus feeders that live on the bottom are food for the fish who swim just above. Brittlestars are no exception, yet some species have a surprise for their predators. When an arm is bitten off, the other arms flash a bright light, startling the predator into retreat. The brittlestar can sacrifice one or two arms in this way and grow more of them, but not many more.

Many deep-sea animals can make light, by the chemical processes known as bioluminescence. The reason may be to confuse predators, attract or illuminate prey,

locate mates of the correct species, or all of these. Movie films of the beautiful pulsing lights produced by delicate combjellies have been taken by submersible craft. A combjelly is a creature with a body like a blob of jelly, which has 'combs' of tiny waving hairs, cilia, arranged in rows along the sides of the body. It uses the cilia to swim, and it catches drifting food with its trailing, sticky tentacles. These animals, also called ctenophores or 'sea gooseberries', resemble jellyfish but are only distantly related.

Some bioluminescent fish have colonies of light-producing bacteria living in their glowing organs, while others make the light themselves, using the chemical luciferin. The deep-sea anglerfish has lights on the end of its spiny fishing-lines, to lure prey into its mouth. Lanternfish have rows of light organs along their sides, like an ocean liner lit up at night. Each species has a different pattern, so they can recognize each other at breeding time.

LEFT: *The deep-sea anglerfish has a light suspended on a 'fishing pole', designed to lure prey into its mouth.*

Fish of the deep sea, such as huge-fanged viperfish, are mostly blind. However, fish that live in the twilight zone, 200–500 m (656–1,640 ft) down, have large eyes specialized to gather as much of the faint light as possible. Hatchetfish have tubular eyes that gather the tiniest glimmer of light reflected off, or produced by, prey fish.

Losses and gains

Because food is so scarce, deep-sea fish have dispensed with all but the most vital body parts; swim bladders, kidneys, brains and muscles are reduced in size, and bones are merely gristle-like struts, since the enormous water pressure supports the body. However, some other body parts are enhanced, to ensure that any food which does come within reach can be caught. Most of these fish have well developed senses of smell and touch, long sharp teeth, and mouths and guts that can expand to swallow items of food much bigger than themselves. The gulper eel is one of the most extreme examples. It is little more than a huge mouth leading to an expandable stomach; head, tail, fins and other body parts are reduced to a bare minimum.

ABOVE: *Bioluminescence may be used to confuse predators, attract prey, or locate mates of the correct species.*

There are other species of fish adapted to the deepest seas, but most of the animals of great depths are invertebrates. Among the most successful are sea-lilies. Millions of them carpet the bottom. They are yet more echinoderms, relatives of the starfish, but compared to the starfish, the sea-lily lives ‘upside-down’, that is, mouth up. It perches on its body stalk on the sea floor, catching drifting food particles with its sticky tube feet.

ABOVE: *Sperm whales are regular visitors to the deeper waters, in search of food such as giant squid.*

Sponges and fanworms also thrive at depth, again by filtering food particles from the slow deep-sea currents. These creatures are generally much smaller, slower-growing and longer-lived than their shallow-water relatives.

Monsters from the depths

Here and there in the abyssal zone, there is a true monster of the deep. The giant Japanese spider crab has a body bigger than a dinner plate, and long spindly claws and legs which span over 4 m (13 ft). It weighs 20 kg (44 lb) and lives by picking edible fragments from the mud. Despite its menacing size, when a giant spider crab is hauled up onto a trawler deck, it can hardly move its limbs, let alone attack or defend itself. The deep water supports its great legs and claws so well, that it has only small muscles to move them. Out of its element, the crab is helpless.

In the world of invertebrates, the giant squid easily beats all records for size. It can grow to 20 m (66 ft) long – from the tip of its body to the ends of its longest tentacles – and weigh 2 tonnes and is a fast and streamlined hunter. It feeds on any fish or other creatures that it can grab with its tentacles, each of which are armed with sharp hooks and powerful suckers. The giant squid also has the largest eyes, the size of a football, of any animal.

Mammals are not designed for deep-sea life. However sperm whales are regular visitors to the deeper waters, on their feeding trips. Indeed, one of their favourite snacks is giant squid. The sperm whale, itself approaching 20 m (66 ft) long, can probably dive over 4 km (2.5 miles). These dives, in water close to freezing point, last over two hours and the whale grabs prey with its stubby, conical teeth. The huge whale returns to the surface, with sucker-shaped scars on its skin and squid in its stomach, as evidence of the great battle in the depths.

Life of plenty, far from the sun

It was thought that huge stretches of the deepest ocean floor could be like deserts – lacking life's essentials, and only thinly populated. So far from the sun, how could animals survive? However the most recent research, by scientists aboard deep-sea submersibles, has revealed that there is a source of energy and nutrition down on the sea floor – from within the Earth itself.

The oases of life are clustered around hot springs, which are rather like underwater volcanoes. Water is heated by the molten rocks beneath the Earth's crust. It spurts up through a vent in the sea floor. Besides hot water, gases and certain chemicals also come from these deep-sea vents, from far below. The chemicals include hydrogen sulphide, a gas which smells like bad eggs, and which is normally noxious to life-forms. Yet some microbes, such as certain sulphur bacteria, thrive on it. They can take in the sulphides and other chemicals, and get energy from them.

These microbes form the basis of food webs around the deep-sea sulphur vents. The animals there have cousins far above, but they are unique species, adapted to the truly extraordinary conditions. Giant tubeworms 3 m (9.75 ft) long, mussels 25 cm (10 in) across, and hundreds of clams and crabs dwell there. The tubeworms and mussels have sulphur bacteria living within their body tissues. The bacteria get the energy and nutrients from the hydrogen sulphide and other chemicals, and share some of them with their hosts, in return for a place to live. The crabs feed on the dead worms and mussels, and the clams filter the general debris.

The deep-sea sulphur-vent communities are the only life on Earth which does not depend ultimately on the Sun.

Unchanging home, unchanged animals

The conditions at the bottom of the sea may be harsh, but at least they are con-

ABOVE: *The marine polychaete worm is microscopic; it is shown here at many times its actual size.*

ABOVE: *Known as the 'living fossil', the coelacanth was thought to have died out millions of years ago, until one was caught in 1938.*

stant. The one factor that these animals do not have to cope with is change, so the deep sea has more than its share of creatures surviving from prehistoric times: they are living fossils, virtually unchanged for millions of years.

The coelacanth is a fish about 1.5 m (5 ft) long which has fins with fleshy bases, or lobes. Coelacanths are known from fossils, but seemed to die out over 70 million years ago. Until, that is, one was caught in 1938, off the coast of south-east Africa. More have been caught since and filmed off the nearby Comoros Islands. Very little is yet known about this strange creature, which swims lazily at depths of a few hundred metres, lurking in wait for food. But its unusual fins give a clue to how, long ago, fish became amphibians and stepped onto land.

Lampshells look rather like clams on stalks. They live in colonies in the deep seas, partly buried in the mud, filtering food from the water. They are named from the resemblance of the shell to an ancient Roman oil lamp. But lampshells are not closely related to clams and other molluscs but are brachiopods, a group of shellfish-type animals. Brachiopod fossils go back almost unchanged for over 500 million years, to a time before anything at all lived on land.

The lampshell probably deserves the title of 'champion living fossil'. It is just one of the amazing creatures hiding in the last great unexplored habitat on Earth – the depths of the oceans.

Index